# Federated Defense in Asia

AUTHORS
Michael J. Green
Kathleen H. Hicks
Zack Cooper

*December 2014*



CSIS | CENTER FOR STRATEGIC & INTERNATIONAL STUDIES

ROWMAN & LITTLEFIELD
Lanham • Boulder • New York • Toronto • Plymouth, UK

## About CSIS

For over 50 years, the Center for Strategic and International Studies (CSIS) has worked to develop solutions to the world's greatest policy challenges. Today, CSIS scholars are providing strategic insights and bipartisan policy solutions to help decisionmakers chart a course toward a better world.

CSIS is a nonprofit organization headquartered in Washington, D.C. The Center's 220 full-time staff and large network of affiliated scholars conduct research and analysis and develop policy initiatives that look into the future and anticipate change.

Founded at the height of the Cold War by David M. Abshire and Admiral Arleigh Burke, CSIS was dedicated to finding ways to sustain American prominence and prosperity as a force for good in the world. Since 1962, CSIS has become one of the world's preeminent international institutions focused on defense and security; regional stability; and transnational challenges ranging from energy and climate to global health and economic integration.

Former U.S. senator Sam Nunn has chaired the CSIS Board of Trustees since 1999. Former deputy secretary of defense John J. Hamre became the Center's president and chief executive officer in 2000.

CSIS does not take specific policy positions; accordingly, all views expressed herein should be understood to be solely those of the author(s).



ISBN: 978-1-4422-4045-2 (pb); 978-1-4422-4046-9 (eBook)

Center for Strategic & International Studies
1616 Rhode Island Avenue, NW
Washington, DC 20036
202-887-0200 | www.csis.org

Rowman & Littlefield
4501 Forbes Boulevard
Lanham, MD 20706
301-459-3366 | www.rowman.com

# Contents

# Acknowledgments

The authors wish to thank the CSIS project staff members for their hard work and valued contributions to this report, particularly John Schaus, T. J. Cipoletti, Gabriel Coll, Samuel Mun, Abby Fanlo, and Lara Crouch. The authors also thank Captain Mark Hooper (USN), Bonnie Glaser, and Rick Rossow for their suggestions and insights on national objectives, capabilities, and potential initiatives.

The authors appreciate the insights of numerous outside experts that attended closed-door and off-the-record CSIS roundtables on federated defense, including those from Capitol Hill, the executive branch, the U.S. military, and foreign governments who contributed their time and opinions to the report.

The project team would also like to thank a host of CSIS's defense industry sponsors. The federated defense project is made possible by their generous annual financial contributions to the Center, which allows us to pursue a wide range of innovative studies. A number of representatives from these companies participated in our workshops and contributed defense industry perspectives to the project team's proposed initiatives, while respecting our intellectual independence at each step along the way.

Although the contributions of all those mentioned above have enhanced the final product, all findings, opinions, and omissions remain those of the authors.

# Executive Summary

The United States has long emphasized the desirability of working with allies and partners to meet pressing security challenges. Indeed, many of our most vexing security concerns—from terrorism to cyber attacks—are best met with concerted multilateral responses. At a time when the United States and many of its allies and partners are reluctant to increase defense and security spending, working together is paramount. This is perhaps most evident in Asia, where present and potential future threats to security and prosperity are high and shared interests are substantial.

The Center for Strategic and International Studies' Federated Defense Project aims to shift our paradigm with key allies and partners from capacity building to federated defense. A federated approach would expand regional security and prosperity by joining regional allies and partners together in the pursuit of shared security objectives across the conflict spectrum. Federated defense should include forward-thinking strategies for how to develop and share capabilities and capacity, thereby more deeply integrating the U.S. military with its allies and partners. This multiyear effort is examining common security goals and the global and regional security architectures and defense capabilities that support them. The project not only focuses on regional approaches but also includes assessments of U.S. foreign military sales, defense acquisition processes, export controls, and other statutory authorities.

This report on federated defense in Asia is the first regional study in the federated defense series. Asia's importance to the United States is clear; as President Obama noted in Canberra, Australia in November 2011, "The United States has been, and always will be, a Pacific nation." Since the end of World War II, U.S. diplomatic, economic, and military engagement have underwritten regional stability and prosperity. U.S. alliances with Japan, South Korea, Australia, the Philippines, and Thailand have formed the basic framework for regional security. In addition, emerging partners, such as India, Indonesia, Vietnam, Malaysia, Myanmar, and others, have helped strengthen regional security. Together with the development of regional institutions, these cooperative efforts have enhanced security and prosperity despite Asia's rapidly changing geopolitical environment.

Today, however, new challenges require new approaches. China's reemergence as a great power is altering security relationships throughout the region, presenting opportunities for enhanced cooperation but also driving military spending and increasing geopolitical tensions. North Korea's provocative actions and its pursuit of nuclear and ballistic missile technology present a continuing security threat, particularly in Northeast Asia. Russia's actions in Ukraine have damaged relations with the United States, as well as many countries in Asia. Meanwhile, regional military investments raise the potential for arms races, worsening the security dilemma throughout the region. Aside from these state-based challenges, there exist numerous non-state threats, such as terrorism, maritime piracy and other illicit activities, and natural disasters. These challenges to the region's security and prosperity require a concerted and coordinated response.

The CSIS federated defense in Asia project was codirected by Dr. Michael Green, senior vice president for Asia and Japan Chair, and Dr. Kathleen Hicks, senior vice president, Henry A. Kissinger Chair, and director of the International Security Program. They led CSIS's project staff in conducting research and holding a series of meetings and workshops designed to review the regional security environment, identify shared regional security objectives, assess common mission areas, expose capability gaps and seams, and evaluate potential federated initiatives. After identifying national objectives and common mission areas, the project team assessed mission requirements on a country-by-country basis. Through a series of workshops and discussions with regional experts, the project team identified existing challenges and opportunities in each mission area. In particular, the project team found critical capability and capacity gaps in six important areas that are ripe for federated approaches. The project team highlighted potential federated efforts based on the geostrategic importance, affordability, and executability of initiatives in these areas.

- *Humanitarian Assistance and Disaster Relief:* Recent natural disasters have demonstrated the need for improved regional collaboration on humanitarian assistance and disaster relief (HA/DR). Regional states could establish additional pre-positioned stockpiles of critical supplies and broaden multinational exercises to include both whole-of-government and nongovernmental cooperation in responding to humanitarian crises.

- *Information and Intelligence Sharing:* The MH-370 disaster heightened awareness of the need for a regional capability to monitor shared air and maritime areas. The United States and its European partners could share insights from efforts to build shared air traffic control and maritime domain awareness architectures in Europe and North America. ASEAN Defense Ministers expressed interest in these mechanisms in early 2014.

- *Maritime Security:* As threats from piracy, illicit trade, transnational crime, and territorial disputes grow, many regional states are recognizing the need for additional maritime sensors and platforms. The cost of acquiring these capabilities is high, but pooled procurement programs for coastal patrol craft and advanced sensors could increase the quantity and quality of maritime assets while creating incentives for interoperability.

- *Undersea Warfare:* The rapid improvement in Chinese and North Korean anti-access capabilities (particularly submarines) puts a premium on undersea operations and anti-submarine warfare. The United States and its allies have long-standing expertise in these areas. U.S. allies and partners could take advantage of this proficiency by working together to field undersea platforms, acquire anti-submarine warfare systems, and improve training and exercising for undersea warfare.

- *Missile Defense:* As cruise and ballistic missiles proliferate throughout Asia, the need for missile defenses is growing. The cost of kinetic interceptors is prohibitive for smaller states and presents a disadvantageous cost-exchange asymmetry for all states. To offset these costs, regional states could share some of the development costs of directed energy and railgun research, which could protect against emerging missile technologies.

- *Cybersecurity:* Although cyber threats have multiplied in recent years, many regional states continue to have limited national capability for cyber operations. As part of the U.S. effort to develop capabilities, operational concepts, and plans to assure access throughout Asia, the United States should coordinate combined cybersecurity exercises to highlight the importance of spending on and cooperation in the cyber domain.

In pursuing these and other federated defense initiatives, officials in the United States and in the region must work together closely with civilian and military leaders, legislative supporters, and defense industry. Such cooperation could help to ensure that regional states not only identify common security objectives, but also work together to meet shared security requirements. This type of federated approach is vital to developing and integrating the region's security capabilities, thereby reinforcing security and prosperity not only within Asia but beyond.

# 1 Introduction

This study examines the potential for "federated defense" in Asia. Federated defense involves integrating U.S., ally, and partner capabilities into regional security architectures that advance common interests. Codirected by Dr. Michael Green, senior vice president for Asia and Japan Chair, and Dr. Kathleen Hicks, senior vice president, Henry A. Kissinger Chair, and director of the International Security Program, this report seeks to explain why federated approaches are needed in Asia and to describe how they might be applied.

The urgency of adopting federated approaches is growing as security threats multiply throughout the region. Although the United States has long emphasized the desirability of working with allies and partners to meet pressing security challenges, deep integration of defense industries and capabilities remains an unachieved objective. However, at a time when the United States and many of its allies and partners are reluctant to increase defense and security spending, working together is of paramount importance. It is time, therefore, to shift our paradigm with key allies and partners from building capacity to federating defense. A federated approach, including forward-thinking strategies for how to develop and share capabilities and capacity, can knit together an Asian security community that is searching for deeper regional integration.

This report makes clear that federated defense can build on existing alliances but is not limited to allies. Indeed, for best effect, federated approaches would connect allies and partners with one another, often with the United States in the background. It is also distinct from an integrated approach because it does not seek to create interdependencies that would impair autonomous action. Instead, by sharing development of federated systems, federated partner countries can draw closer to one another and to the United States through training, logistical support, tactical development, and, potentially, operational missions. By better leveraging select host nation facilities, federated efforts could maintain the "low-cost, small-footprint" approach that is both affordable and suited to regional dynamics. This concept aims to be both strategically significant and cost-effective, building on the natural desire of allies and partners for closer working ties with one another and with the United States.

CSIS's broader federated defense project is a multiyear effort examining common security goals and the global and regional architectures and defense capabilities that best support them. This project aims to take advantage of the growth in global value chains that

The Wasp-class amphibious assault ship USS *Bonhomme Richard*, center; the Republic of Korea Navy guided-missile destroyer *Sejong the Great*, left; the Japanese Maritime Self-Defense Force destroyer JS *Akebono* and other ships assigned to the Rim of the Pacific (RIMPAC) 2010 combined task force transit the Pacific Ocean in a 32-vessel formation north of Hawaii on July 24, 2010. RIMPAC, the world's largest multinational maritime exercise, is a biennial event that allows participants to work together to build trust and enhance partnerships needed to improve maritime security. Reprinted with permission from the U.S. Department of Defense.

have reshaped the commercial defense sector worldwide. Falling trade costs and lowered barriers to the movement of ideas and expertise are encouraging global innovation. With the dispersion and diffusion of both production and innovation, the United States will have to rely on closer cooperation with allies and partners to maintain its technological edge and sustain a robust defense industrial base. With the growth of Asian defense industry, seizing opportunities for collaborative efforts is vital.

This report on federated defense in Asia will be followed by other regional studies, including the Middle East and Europe. The project is also assessing U.S. foreign military sales, defense acquisition processes, export controls, and other statutory authorities. These efforts draw on the full breadth of CSIS's expertise, ranging from scholars with deep regional knowledge and experience to former government officials involved in development of U.S. and regional defense concepts, capabilities, posture, and relationships to experts on economics, trade, and global defense industry.

In conducting the Asia assessment, CSIS project staff proceeded in four phases. The first phase validated the regional federated defense construct. The second phase assessed objectives, priorities, and potential initiatives in South and Southeast Asia. The third phase validated objectives, priorities, and potential initiatives in Northeast Asia. The final phase of the project included the selection of initiatives and the release of this report. Throughout this project, the CSIS project team conducted and held a series of meetings and workshops to gain insight from experts in the executive and legislative branches of the U.S. government, foreign government officials, industry representatives, and regional security scholars.

The sections that follow address each of the issues raised in the study effort. Section 2 briefly assesses regional security challenges and opportunities in Asia. Section 3 identifies common security objectives, mission areas, and capability gaps. Section 4 reviews six specific, concrete, and tractable proposals for potential federated initiatives, selected based on the above analysis and the need for financially affordable and realistically executable initiatives. The final section concludes by noting the importance of involving a range of domestic and international partners—including foreign leaders as well as those in the U.S. executive branch, the Congress, the U.S. military, and the global defense industry—if federated approaches are to be adopted.

# 2 Regional Security Challenges and Opportunities

Nowhere around the globe are new federated initiatives more promising and existing collective action mechanisms less tested than in Asia, where current and potential future threats to security are substantial and shared interests are abundant.[1] The Asia-Pacific region is the world's most dynamic security environment and is home to its three largest economies (the United States, China, and Japan); three most populous states (China, India, and the United States); and nearly two-thirds of global defense spending (when Russia is included).[2] As China's economic, military, and geopolitical influence grows, the region is witnessing the largest shift in the global distribution of power since the United States rose in the late nineteenth and early twentieth centuries. These changing circumstances necessitate new approaches and arrangements to protect regional security.

As President Obama stated when he announced the U.S. rebalance to Asia in November 2011, "the United States has been, and always will be, a Pacific nation."[3] Since the end of World War II, U.S. diplomatic, economic, and military engagement have underwritten regional stability and prosperity. U.S. alliances with Japan, South Korea, Australia, the Philippines, and Thailand have formed the basic framework for regional security. In addition, partners such as India, Indonesia, Malaysia, Singapore, Taiwan, and Vietnam, among others, have helped to strengthen regional security. Together with the development of regional institutions, such as the ASEAN Regional Forum (ARF), Asia-Pacific Economic Cooperation (APEC), and ASEAN Defense Ministers Meeting-Plus (ADMM+), these cooperative efforts have enhanced security and prosperity despite Asia's rapidly changing geopolitical environment.

Today, however, security challenges are multiplying across the region. China's reemergence as a great power is altering security relationships throughout Asia, presenting opportunities for enhanced cooperation—including with the United States—but also

1. This study focuses primarily on states in maritime Asia, rather than Central Asia, because of the increased potential for multilateral cooperation on shared maritime issues.

2. Asia, including Russia and the United States, accounts for 63.5 percent of global defense spending, according to the International Institute for Strategic Studies (IISS), *The Military Balance 2014* (London: Routledge, 2014), 485–492.

3. Barack Obama, "Remarks to the Australian Parliament, The White House, November 17, 2011," http://www.whitehouse.gov/the-press-office/2011/11/17/remarks-president-obama-australian-parliament.

driving new military spending and geopolitical realignment. China's continued rise is by no means assured, but Xi Jinping's rapid consolidation of power since assuming the role of president and general secretary of the Chinese Communist Party indicates that China's leaders may take a more active approach to solving both domestic and international challenges. In recent years, China's growing assertiveness in maritime disputes has inflamed tensions with its neighbors in the East and South China Seas. The United States and its allies and partners in the region will need an approach that dissuades coercion in these territorial disputes while building patterns of cooperation and confidence with China going forward.

China's rise is certainly the most dramatic change in Asia, but it is not the only state-based challenge to regional security. Another danger stems from North Korea's continuing security threat to its neighbors in Northeast Asia. North Korean provocations remain a concern, such as the shelling of Yeonpyeong Island, the sinking of the South Korean corvette *Cheonan*, North Korea's unfettered nuclear and missile programs, and the risk that the regime may pursue provocative and coercive approaches with increased impunity. The Korean peninsula has grown more unstable since Kim Jong-un succeeded his father as supreme leader, as evidenced by North Korea's continued violations of United Nations Security Council resolutions and by the execution of Kim's powerful uncle Jang Sung-taek. Further to the west, Russian president Vladimir Putin's actions in Ukraine have badly damaged relations between Russia and Europe and the United States, as well as many countries in Asia. Russia can play an important and positive role as an energy supplier to East Asian states and as a member of the Six-Party Talks with North Korea. However, Russia's actions in Europe and its increased operational tempo in the Far East raise questions about Russia's potential as a partner and its commitment to uphold existing international rules and norms.

The rapid growth of other regional militaries also presents the potential for arms races and the worsening of the security dilemma throughout the region. Japan's more proactive security engagement is essential for regional security, but such efforts must be clearly explained to, and welcomed by, its neighbors, lest tensions with South Korea, China, and Taiwan grow. India's growth also has the potential to alter regional dynamics, particularly as India confronts Chinese incursions in disputed border areas and pushes its naval forces farther into the Indian Ocean. Indeed, across Southeast Asia, regional states are investing in their militaries. As these states build new capabilities, it will be increasingly important that they work together to enhance regional security and avoid the potential for destabilizing arms races.

Aside from these state-based challenges, there exist numerous non-state threats in Asia. Foremost among these is the danger of transnational terrorism, a constant challenge for governments in India, the Philippines, Indonesia, Malaysia, Singapore, China, and elsewhere. Maritime piracy and illicit activities such as trafficking in drugs and people also require enhanced regional cooperation, lest local insecurity spread. In addition, natural disasters remain a serious threat to regional security and prosperity, as evidenced by Typhoon Haiyan in the Philippines, the Great East Japan Earthquake and Tsunami, and the Indian Ocean Earthquake and Tsunami. Other man-made disasters, such as the disappearance of Malaysia Airways flight 370, also highlight the importance of deepening multilateral cooperation to address nontraditional security challenges.

Armed Forces of the Philippines service members take a break from moving humanitarian supplies in Iloilo, Philippines. The *Ronald Reagan* carrier strike group relocated off the coast of the Philippines to provide HA/DR to victims of Typhoon Fengshen, July 1, 2008. Reprinted with permission from the U.S. Department of Defense.

# 3 Common Security Objectives, Missions, and Capabilities

Finding ways to navigate these growing challenges is critical to regional security and prosperity. Establishing a sustainable regional security architecture will require that states identify common security objectives and then work to meet those goals in a cooperative manner. This project therefore began with an assessment of national strategy documents, senior leader statements, and expert analyses to identify common security objectives. This section then continues with an assessment of mission areas common across many regional states and concludes with an evaluation of critical gaps and seams in existing capabilities.

## Common Security Objectives

Five security objectives were largely consistent among America's allies and partners in the region: protecting territorial integrity, avoiding the emergence of a hostile regionally dominant state, ensuring the free flow of commerce, preventing proliferation of weapons of mass destruction, and supporting the rule of law, both domestically and internationally. These five security objectives drive regional militaries' missions and requirements. The project team reviewed the top military missions of the major regional states to identify their security objectives. In analyzing common mission areas, a number of shared priorities emerged, spanning all domains of warfare and all intensities of conflict. Some mission areas were common across every country, such as humanitarian assistance and disaster relief, information and intelligence sharing, and cybersecurity. Cooperative efforts in response to non-state challenges have the potential to improve transparency and confidence building among states that sometimes are in tension over territorial issues. Failure to address these challenges can also lead to instability among weaker states and foster competitive regional dynamics. The importance of other mission areas varied substantially across the region, particularly those missions, such as missile defense, necessary for deterring or defeating potential state-based threats to disputed territory. Although there has long been more attention paid to state-based threats in Northeast Asia than in Southeast Asia, the project team found that Southeast Asian states are increasingly focused on state-based challenges, despite their more limited investments in defensive capabilities.

**Figure 1. Difficulty of Federating Capabilities across Asia**

# Critical Mission Areas

Figure 1 seeks to describe the types of military missions pursued across Asia, from the most cooperative to the most competitive. The most cooperative missions—those which countries most readily engaged in with others in the region—typically included positive-sum cooperative efforts such as HA/DR, counterterrorism, and coastal patrol, each of which appeared in most national strategy documents across the region. Many states in Southeast Asia are likely to focus on these missions given their less technologically demanding requirements. A smaller subset of countries is engaging in more competitive military mission areas, such as air and missile defense, undersea warfare, and counter–anti-access/area denial (A2/AD). These missions were mostly commonly supported by Northeast Asian states facing heightened state-based threats from China and North Korea, as well as the technologically advanced militaries of India, Singapore, and Australia. A third set of capabilities that might be termed cross-cutting enablers—including cybersecurity, information sharing, and intelligence, surveillance, and reconnaissance (ISR) collection—was emphasized across the region.

Although the ability of various states to safeguard shared security objectives differs, overlapping interests are likely to allow for closer cooperation. Therefore, after assessing common mission areas, the CSIS project team assessed mission requirements on a country-by-country basis. The findings of this work are shown in Table 1. Note that the table

## Table 1. Security Missions Being Pursued in Asia[1]

| | *Australia* | *Brunei* | *Cambodia* | *China* | *India* | *Indonesia* | *Japan* | *South Korea* | *Laos* | *Malaysia* | *Myanmar* | *The Philippines* | *Russia* | *Singapore* | *Taiwan* | *Thailand* | *Vietnam* | *United States* | *Other Partners*[1] |
|---|---|---|---|---|---|---|---|---|---|---|---|---|---|---|---|---|---|---|---|
| HA/DR | ✓ | ✓ | ✓ | ✓ | ✓ | ✓ | ✓ | ✓ | ✓ | ✓ | ✓ | ✓ | ✓ | ✓ | ✓ | ✓ | ✓ | ✓ | ✓ |
| Information Sharing | ✓ | ✓ | ✓ | ✓ | ✓ | ✓ | ✓ | ✓ | ✓ | ✓ | ✓ | ✓ | ✓ | ✓ | ✓ | ✓ | ✓ | ✓ | ✓ |
| Peace Operations[2] | ✓ | ✓ | ✓ | ✓ | ✓ | ✓ | ✓ | ✓ | | ✓ | | ✓ | ✓ | ✓ | ✓ | ✓ | | ✓ | ✓ |
| Counter-Terrorism | ✓ | ✓ | ✓ | ✓ | ✓ | ✓ | ✓ | ✓ | ✓ | ✓ | ✓ | ✓ | ✓ | ✓ | ✓ | ✓ | ✓ | ✓ | ✓ |
| Ground Operations | ✓ | ✓ | ✓ | ✓ | ✓ | ✓ | ✓ | ✓ | ✓ | ✓ | ✓ | ✓ | ✓ | ✓ | ✓ | ✓ | ✓ | ✓ | ✓ |
| Coastal Patrol | ✓ | ✓ | | ✓ | ✓ | ✓ | ✓ | ✓ | | ✓ | ✓ | ✓ | ✓ | ✓ | ✓ | ✓ | ✓ | ✓ | ✓ |
| Maritime Domain Awareness | ✓ | ✓ | | ✓ | ✓ | ✓ | ✓ | ✓ | | ✓ | | ✓ | ✓ | ✓ | ✓ | ✓ | ✓ | ✓ | ✓ |
| ISR | ✓ | ✓ | ✓ | ✓ | ✓ | ✓ | ✓ | ✓ | ✓ | ✓ | ✓ | ✓ | ✓ | ✓ | ✓ | ✓ | ✓ | ✓ | ✓ |
| Cyber Security | ✓ | ✓ | ✓ | ✓ | ✓ | ✓ | ✓ | ✓ | ✓ | ✓ | ✓ | ✓ | ✓ | ✓ | ✓ | ✓ | ✓ | ✓ | ✓ |
| Missile Defense | ✓ | | | ✓ | ✓ | | ✓ | ✓ | | | | | ✓ | ✓ | ✓ | | | ✓ | ✓ |
| Air Superiority | ✓ | | ✓ | ✓ | ✓ | ✓ | ✓ | ✓ | | ✓ | | ✓ | ✓ | ✓ | ✓ | ✓ | ✓ | ✓ | ✓ |
| Undersea Warfare | ✓ | | | ✓ | ✓ | ✓ | ✓ | ✓ | | ✓ | | | ✓ | ✓ | ✓ | | ✓ | ✓ | ✓ |
| Expeditionary Warfare | ✓ | | | ✓ | ✓ | | ✓ | ✓ | | | | ✓ | ✓ | | | | | ✓ | ✓ |
| Counter-A2/AD Measures | ✓ | | | | ✓ | | ✓ | ✓ | | | | | | | ✓ | | | ✓ | |

[1] Other partners primarily include U.S. NATO allies, as well as arms suppliers such as Israel, Brazil, and others.
[2] See UN Peacekeeping, "Troop and police contributors," June 2014, http://www.un.org/en/peacekeeping/resources/statistics/contributors.shtml.

1. As noted in the text, Table 1 reflects the expressed desire of countries to carry out missions in certain areas, not those countries' actual capabilities or the likelihood that they might cooperate with the United States.

reflects the expressed desire of countries to carry out certain missions, not those countries' actual capabilities. Moreover, the inclusion of a country in a shared mission area does not necessarily indicate potential partnership with the United States. For example, some areas of U.S. cooperation with Thailand, Vietnam, Myanmar, Russia, and China are restricted by congressional legislation. In addition, the United States is likely to limit the extent of cooperation with potential adversaries in some mission areas, such as missile defense, which are designed to protect against state-based threats.

Assessing shared mission areas is not sufficient to identify the most important areas for federated approaches. To yield the greatest benefit, federated initiatives must be designed to address areas of need, requiring study of regional capability gaps and seams.

## Capability Gaps and Seams

After identifying common missions and the countries most interested in each mission area, the project team assessed existing capability gaps and seams. Through a series of workshops and discussions with regional experts, the project team identified challenges and opportunities in each mission area. Some missions, such as counterterrorism, have garnered substantial attention in recent years from both the United States and from regional states. Although more work is needed to address continuing challenges in these areas, ongoing efforts to identify and address needs have already yielded many results. In other areas, however, there is an urgent need for more federated approaches. The study team identified six such areas on which to focus its efforts.

One notable example is the growing requirement across Asia for maritime domain awareness and coastal patrol capabilities. As maritime trade continues to grow, regional fishing fleets expand and compete over dwindling fish stocks, and both non-state threats and state-based disputes over sovereignty rise, many gaps and seams are opening in the maritime domain. The need to collect, analyze, share, and act on maritime ISR is growing rapidly, outpacing the capabilities of regional states, particularly smaller and less-well-funded coast guards and militaries. In addition, the ability to respond to crises or conflicts with maritime assets, be they coast guard or naval platforms, is particularly limited in Southeast Asia. Therefore, there exists a dangerous and growing capability gap in maritime security, one that federated approaches can help to address.

Other capability areas selected by the CSIS team for study were HA/DR, information and intelligence sharing, undersea warfare, missile defense, and cybersecurity. There are substantial capability gaps in each of these capability areas and a coalescence of interests among at least some partners and allies, making these areas ripe for federated approaches that quickly and economically build regional capability and capacity.

# 4 Potential Federated Initiatives

After identifying these six critical capabilities, the project team reviewed potential initiatives that might help to address each. Potential initiatives were culled by a combination of prior research findings, existing collaborations within Asia, cooperative successes in other geographic areas, and input from regional and functional experts both in the United States and throughout the region.

## Selection Criteria for Potential Initiatives

The project team examined dozens of potential initiatives but selected only the most promising based on the following criteria:

- *Geostrategic Importance:* The extent to which the initiative would improve relations among Asian allies and partners, deter provocative actions, and shape strategic behavior. Those initiatives with the highest geostrategic importance often require the largest investments and overcoming significant political barriers to yield results. For example, developing intelligence-sharing systems would be complicated and costly for all countries involved, but it would greatly expand the collective capability of regional states.
- *Affordability:* The extent to which implementation and sustainment costs to the United States and regional states differ from status quo policies. In assessing affordability, the project team differentiated between those states with high or growing defense expenditures and those states with more limited defense spending levels. Sustainment costs, which are often overlooked in the acquisition process, are particularly critical for states with more limited budgets or less-developed maintenance and operational experience.
- *Executability:* The extent to which the initiative is feasible and can be implemented and sustained within desired time frames. Within this construct, executability includes both political feasibility and technical capability. For example, despite the potential benefits of operating nuclear submarines, Australian leaders have decided that political support, technical capacity, manpower requirements, and industrial expertise are lacking in Australia. As a result, diesel-powered submarines have been identified by Australia as a more attractive option.

On November 17, 2009, the Seawolf-class attack submarine USS *Connecticut* underway in the Pacific Ocean with an HH-60H *Sea Hawk* helicopter from the Chargers of Helicopter Anti-Submarine Squadron 14. In the background are the aircraft carrier USS *George Washington* and the Japan Maritime Self-Defense Force helicopter destroyer JS *Hyuga*. Ships from the U.S. Navy and Japan Maritime Self-Defense Force are participating in Annual Exercise, a bilateral exercise designed to enhance the capabilities of both naval forces. U.S. Navy photo by Mass Communication Specialist 1st Class John M. Hageman, released for publication.

Described below are six federated initiatives that are highly important, relatively affordable, and realistically executable. Each illustrative example demonstrates ways that federated approaches could benefit regional security and prosperity. These six initiatives are intended to demonstrate that federated approaches have potential in a wide variety of areas, from low- to high-threat environments, from cooperative to competitive scenarios, from relatively low-technology solutions to highly demanding and innovative areas, and across all domains of warfare.

On October 1, 2014, U.S. Marines with Battalion Landing Team, 3rd Battalion, 5th Marines, assigned to the 31st Marines Expeditionary Unit carry a combat rubber raiding craft onto the shore to set up a simulated assault during Amphibious Landing Exercise (PHIBLEX) 15 in the Philippines. PHIBLEX is an annual, bilateral training exercise conducted by the Armed Forces of the Philippines, U.S. Marines and Navy to strengthen interoperability across a range of capabilities to include disaster relief and contingency operations. (U.S. Marine Corps photo by Pfc. Matthew Casbarro).

# Humanitarian Assistance and Disaster Relief

Among the most important human security priorities in Asia is improving capability and capacity for rapid delivery of HA/DR. Recent natural disasters, such as Typhoon Haiyan, the Great East Japan Earthquake and Tsunami, and the Indian Ocean Earthquake and Tsunami, have demonstrated the need for improved regional collaboration on HA/DR missions. The proximity of many highly populated Asian states to areas frequently affected by earthquakes, tsunamis, typhoons, and floods creates an urgent need for better coordination. These requirements are likely to grow as the hazards of climate change and rising sea levels increase the potential for large-scale natural disasters.

Despite widespread recognition of the importance of HA/DR there remain serious limitations in regional response capacity and capability. Lack of sufficient stockpiles of food, water, and other emergency supplies has slowed previous HA/DR efforts. Although there are major stockpiles in Subang, Malaysia, Brisbane, Australia, and Singapore, these stockpiles are insufficient to address the massive potential need in a widespread disaster.[1]

Stockpile shortages have been particularly problematic where local transportation networks have been damaged, further slowing delivery of emergency supplies. Transporting needed supplies from stockpiles to affected individuals, particularly in outlying areas, is especially difficult. Large-capacity transport aircraft and military vessels are ideal for transporting aid into affected areas, but local delivery requires other means. Efforts to address these deficiencies through acquisition of more light transport vehicles could make a difference, but transport helicopters are likely to be a high-demand, low-density asset in a crisis.

In addition, the challenges of coordinating actions across multiple agencies, governments, and nongovernment organizations have proven difficult without prior cooperation. Inexperienced personnel and limited interagency and nongovernmental cooperation remain problematic. Most concerning, lack of prior coordination has slowed response efforts, even when capability and capacity have existed. Efforts by regional states to broaden multinational exercises to include whole-of-government and nongovernmental cooperation are badly needed. Some steps have already been taken, including efforts by the ASEAN Regional Forum to include both civilian and military personnel in its biennial Disaster Relief Exercise (DiREx). Recent exercises have shown the potential of this type of coordination, but they have also highlighted the challenges of coordinating civilian and military organizations, even within a single government.

1. Jennifer D. P. Moroney et al., *Lessons from Department of Defense Disaster Relief Efforts in the Asia-Pacific Region* (Santa Monica, CA: RAND, 2013), http://www.rand.org/content/dam/rand/pubs/research_reports/RR100/RR146/RAND_RR146.pdf.

**Illustrative Example: HA/DR Stockpiles, Transportation, and Exercises**

Recent HA/DR efforts have highlighted the challenge of rapidly delivering humanitarian supplies to affected populations. Three challenges have been particularly difficult: (1) locating and accessing stockpiles of needed supplies, (2) transporting those goods to affected areas, and (3) coordinating multinational and nongovernmental efforts. Creating a more federated HA/DR network could help to increase the speed and effectiveness of HA/DR efforts.

The requirements of widespread natural (and sometimes man-made) disasters surpass the capabilities of any one state, but a multinational effort to increase both the size of these stockpiles and their geographic dispersal could increase the speed of regional responses. Such efforts should be focused on establishing stockpiles near major centrally located hubs, such as U-Tapao in Thailand or other similarly situated locations throughout Southeast Asia that have the capacity for large logistical burdens.

A more federated system might also focus on increasing local transport capabilities. As nations throughout the region have found, the challenge of providing critical humanitarian assistance immediately after a crisis is often worsened by damage to local transportation networks. In the aftermath of a natural disaster, rotary-wing aircraft with large payloads are especially important. Yet these assets are relatively scarce and in high demand, particularly in Southeast Asia. Therefore, regional states should consider joint procurement of utility and transport helicopters to increase regional capacity and drive down unit costs, similar to NATO's airborne early warning and control arrangement. Such efforts would also improve interoperability during crises. An alternative form of joint procurement would be an established, multinational rotational "on call" force of medium and heavy-lift aircraft, similar to the NATO Response Force. Regional militaries could volunteer for several-month rotations where their assets would be "on call" to assist in natural disasters. Countries could build this "on call" period into training and deployment cycles, and simply be available if necessary.

Finally, efforts to locate new stockpiles and acquire or rotate rapid delivery capabilities must be exercised before they are used. These exercises should include not just regional militaries but also nonsecurity agencies and nongovernmental organizations. The United States already conducts some exercises that incorporate first responders and nongovernmental organizations within the United States, so broadening these efforts could help to underscore the reality that frequent training and exercising is essential to proper execution during a crisis. Incorporating a broad range of potential partners, including China, into such exercises could help facilitate a large and federated HA/DR network.

## Information and Intelligence Sharing

The MH-370 disaster heightened awareness of the need for a concerted regional effort to monitor air and maritime areas in Southeast Asia. Limitations in air and maritime surveillance and information sharing are a challenge not only in Southeast Asia but elsewhere in Asia as well. With overlapping territorial claims, many small private vessels, and bustling commercial choke points, the need for shared operating pictures is likely to grow in future years. Despite this requirement and the reality that all states in the region have an interest in improving access to information and intelligence, regional states have been hesitant to share the information to which they have access.

Intelligence sharing among allies must remain a top priority. Although U.S. information- and intelligence-sharing architectures are robust, many links between U.S. allies remain underdeveloped. Japan and South Korea, for example, would benefit greatly from conclusion of a General Security of Military Information Agreement (GSOMIA), which would allow them to more seamlessly share critical information and intelligence on security threats. Japan's new information security legislation and South Korea's operational control transition arrangements provide an opportunity to address this gap. Trilateral arrangements, potentially involving Japan, South Korea, and Australia, would further expand information and intelligence-sharing networks.

Although technical challenges persist, the main limiting factor in information and intelligence sharing among allies, partners, and competitors is the willingness to share. Despite decades of cooperation between ASEAN member states, cultural and political barriers to cooperation remain. Of critical importance, therefore, is persuading both senior leaders and lower-level government officials that sharing is in their interests. Rational reasons to restrict sharing of some information, and in particular intelligence, will remain, but cultural and political challenges can and should be overcome.

History shows that the most progress has been made when sharing information and intelligence is vital to continued security. In Northeast Asia, for example, efforts are ongoing to share radar data required for regional missile defenses. Taiwan's advanced radars, linked with South Korean and Japanese missile defense capabilities, have helped to provide a strengthened missile defense architecture, particularly useful for tracking and potentially intercepting North Korean ballistic missiles. This contrasts with information-sharing efforts in Southeast Asia, which have generally failed to bridge cultural, political, and trust gaps. Although Singapore's Changi Information Fusion Centre brings together representatives from throughout Asia to share information and intelligence, it has been underutilized and hampered by a lack of trust.

One area where information sharing has occurred in Southeast Asia is the Malacca Strait Patrols. Today, this cooperative effort includes Indonesia, Malaysia, Singapore, and Thailand, which share information on maritime sea-lanes. Multiple governments cooperate through Eyes-in-the-Sky aerial surveillance patrols, the information-sharing Intelligence Exchange Group, and routine coordinated patrols and the Monitoring and Action

Agency. Coordination has decreased incidents of piracy in this vital commercial area. Nevertheless, resistance to sharing has prevented architectures like the Malacca Straits Patrol from being applied more broadly.

**Illustrative Example: Extraregional Information-Sharing Mechanisms**

Although information sharing is critical to regional security cooperation, it has proven difficult to accomplish given national concerns and sensitivities. However, with increasingly complex regional challenges, particularly in shared maritime areas, the need to adopt information- and intelligence-sharing measures is growing more urgent.

The United States and its European allies and partners have experience working with each other outside Asia that might be applicable inside Asia. At an April 2014 meeting of ASEAN Defense Ministers in Hawaii, Admiral Samuel Locklear, commander of U.S. Pacific Command, suggested that European arrangements to share information and intelligence with allies and partners might prove to be a useful guide. Regional leaders have already expressed interest in these multilateral information collection and sharing architectures, making this an ideal area for future cooperation. Although technical challenges will remain, these examples could help to ease concerns over trust issues and provide a guide to sharing sensitive data.

The United States could facilitate the sharing of lessons learned from U.S. and European efforts to construct sharing architectures in North America and Europe. This would provide a point of entry for the European Union and its member states to offer meaningful expertise and resources to Asian partners. Some specific initiatives of interest are NATO's Operation Active Endeavor, the EUROCONTROL air traffic management system, Baltic Sea cooperation, and the pan-European Border Surveillance System. NATO's Operation Active Endeavor monitors shipping in the Mediterranean and includes not only NATO countries but other partners as well. EUROCONTROL operates the first international air traffic control center, providing an air traffic management system expected to handle more than 50,000 flights by 2020. Sea Surveillance Cooperation Baltic Sea involves various bilateral sharing initiatives between Finland and Sweden. Finally, the pan-European Border Surveillance System incorporates various information-sharing mechanisms into a larger regional architecture.

These types of information-sharing efforts should build on broad-based open-source architectures. Open systems foster opportunities to increase information flows, creating patterns of cooperation on which to build further shared knowledge and capabilities. Open-source efforts could allow the inclusion of countries, such as China, that might not otherwise take part in information-sharing efforts, thereby helping to build a shared picture of air and maritime activities across the region.

# Maritime Security

Asia's seaborne trade continues to boom, with half of the world's merchant fleet tonnage passing through the Strait of Malacca, Sunda Strait, and Lombok Strait.[2] Furthermore, more than one-third of global crude oil and more than half of global liquefied natural gas passes through the South China Sea, figures that are expected to rise as Asian economies grow. In addition, fishing and extraction of oil and natural gas resources make maritime regions a vital source of sustainable economic wealth.

Asia's dependence on maritime trade and commerce requires a robust regional partnership to protect maritime security. As already discussed, regional cooperation on maritime issues is substantial, and it includes the Regional Cooperation Agreement on Combating Piracy and Armed Robbery against Ships in Asia (ReCAAP), the Malacca Strait Patrol, and the Declaration on the Conduct of Parties in the South China Sea. Yet, facing continued threats from piracy, illicit trade, and transnational crime, many regional states, particularly in Southeast Asia, are recognizing the need for additional efforts to address shared challenges. Reinforced by concerns over territorial disputes in the South and East China Seas, these states seek to increase the quality and quantity of their maritime platforms and sensors.

Federated approaches to maritime security issues are particularly attractive because many maritime threats are shared and many of the capabilities required are similar. Although the cost of developing new sensors and platforms is high, particularly for cash-strapped regional coast guards and navies, joint development can help to ease the burden. A common need across many states is for additional coastal patrol vessels, often operated by fisheries agencies or coast guards rather than navies. With increasingly crowded sea-lanes and overlapping territorial claims, such cooperative efforts are critical for increasing regional security and prosperity.

Some bilateral cooperation on maritime platforms has already occurred. Many regional states already build relatively affordable patrol craft, corvettes, and frigates. Japan, for example, has strengthened its ties with both the Philippines and Vietnam, arranging transfers of new and used patrol craft to its South China Sea partners. The United States also has a role to play. U.S. leaders have agreed to transfer more than $150 million in funding for maritime capacity building in Southeast Asia in coming years, to include coast guard vessels for Vietnam. Indonesia, Malaysia, Thailand, and others in the region also face severe shortages of coastal patrol craft and could benefit from additional security assistance.

In addition to the patrol craft themselves, sensors and information sharing are also critical. Advanced radars are included in some naval vessels, but many coast guard and smaller naval ships lack the requisite intelligence, surveillance, reconnaissance, and

---

2. Data from U.S. Energy Information Agency, "South China Sea," February 2013, http://www.eia.gov/countries/analysisbriefs/South_China_Sea/south_china_sea.pdf.

communications capabilities. Regional partners could help to close these gaps. Australia, for example, already provides maritime domain awareness information to Malaysia through the Five Powers Defense Arrangement. Engagements of this sort can increase the effectiveness of patrol vessels, targeting illicit activity, aiding in search and rescue missions, and preventing disputes over operational areas. As discussed in the preceding section on information and intelligence sharing, ensuring that information exchanges are robust will also enable cooperation and action by all participating countries.

**Illustrative Example: Regional Cooperation on Coastal Patrol Craft**

From the East China Sea to the Indian Ocean, the need for coastal patrol craft is growing rapidly. These maritime vessels can help monitor disputed territories and protect vital trading routes. Enabled by ISR assets as well as regional information-sharing efforts, these coastal patrol craft could dramatically increase maritime security from the Indian Ocean to the Pacific. Yet, for many smaller maritime states, the cost of modern coastal patrol craft and their larger crew sizes is prohibitive, particularly when combined with the need to integrate advanced sensor and communications systems.

Regional cooperation on coastal patrol craft could increase the number of available patrol vessels while improving interoperability. Regional states (rather than the United States) are likely to be the most attractive partners, given their ability to produce low-cost but high-quality naval systems. Singapore, South Korea, and Japan have highly capable shipbuilding industries and substantial experience building coast guard and naval platforms. In addition, these states field advanced radars and sensor systems, some of which would be valuable not only to regional navies but to coast guards, police agencies, and fisheries administrations, as well. Other regional states may be able to provide relatively low-cost components for these vessels, particularly given the lower cost of labor in many parts of Southeast Asia. The United States could use its expertise in training and maintenance to help ensure that allies and partners maximize the capability of their new and existing systems. This building partner capacity experience in training foreign militaries to operate and maintain complex systems can help to supplement allies' and partners' efforts to increase the quality and quantity of maritime security vessels, particularly in Southeast Asia.

## Undersea Warfare

Given their ability to operate in nonpermissive areas, attack submarines are in high demand throughout Asia, especially as maritime tensions rise. The growth in Asia's submarine fleets, led by China's 136 submarines, has been a concern for a number of countries. Whereas large regional economies such as China, Japan, and South Korea have developed indigenous submarine models, many regional states have acquired new undersea systems from abroad, such as Vietnam's procurement of *Kilo*-class submarines from Russia and Malaysia's purchase of a *Scorpène*-class submarine. The high cost of submarines and antisubmarine warfare capabilities, however, puts a heavy burden on defense spending and simultaneously increases the potential for regional arms races. A more federated system might allow regional states to field undersea warfare capabilities at a more reasonable cost and limit their effect on the regional security dilemma.

With the exception of India, which operates Russian *Akula*-class nuclear submarines, and China, no regional states field nuclear attack submarines, limiting the options for quiet and long-endurance submarines to diesel boats. The United States also does not produce conventionally powered attack submarines, but many U.S. allies have a long history of submarine production. Japan's *Sōryū*-class, for example, is one of the world's best air-independent propulsion submarines. Although the *Sōryū*-class has never been exported, revision of Japan's three principles of arms export may allow production for foreign partners. In addition, South Korea's *Type 214* submarine also uses diesel-electric propulsion and is already used by several European partners.

Japanese and South Korean submarines may be attractive to partners throughout Asia who have typically developed submarines indigenously or procured them from Europe or Russia. Australia, for example, has struggled with reliability and maintenance problems with its indigenous *Collins*-class submarines. Taiwan has also experienced difficulties operating its *Hai Lung*-class. Indonesia has purchased submarines from Germany; India and Malaysia from French and Spanish builders; Singapore from Sweden; and Vietnam and India from Russia. Most of these submarines, however, tend to be far smaller and have less endurance and reduced payloads compared with the *Sōryū* or *Type 214* submarines.

In addition to submarine production, regional states could improve antisubmarine warfare capabilities to counter the growing proliferation of regional submarines. Once again, Japan has long-standing expertise in this area, and both Japan and the United States field advanced anti-submarine warfare systems that should be attractive to regional partners. The U.S.-built P-8 *Poseidon*, which India and Australia have already purchased, is specifically designed for long-range maritime patrols, with capability against both surface and subsurface threats. Japan's P-1 is also a highly capable maritime patrol aircraft, providing both a sensor and weapons platform. Rotary-wing aircraft, such as the SH-60 *Seahawk* helicopter operated by the United States, Australia, Japan, Singapore, Taiwan, and Thailand, also provide substantial capability. Regional procurement of similar systems would aid interoperability.

Allies and partners could take advantage of these anti-submarine warfare capabilities, as well as advanced sensor networks, surface naval platforms, and unmanned underwater vehicles, not only by procuring new platforms but also by sharing operational expertise. Combined training and exercising would help regional states counter undersea threats. Such combined efforts could be aided by forward deployment of U.S. naval assets, to include both submarines and maritime patrol aircraft. This would also increase the time-on-station of high-demand U.S. undersea and maritime patrol assets. These efforts would enhance other maritime cooperation and expand the capability and capacity of the United States and its allies and partners in the undersea domain.

**Illustrative Example: Japanese Cooperation on Australia's Future Submarine**

As Australia begins development of its Future Submarine, cooperation with foreign partners is becoming increasingly attractive. The Royal Australian Navy expects to acquire 10–12 submarines at a cost of $20–$30 billion USD. Although indigenous production would be preferred, some Australian leaders have questioned the Australian defense industry's ability to produce capable submarines at the cost and on the timelines required. Australia's existing *Collins*-class submarines have suffered from poor reliability, limiting their operational effectiveness and adding motivation to seek outside expertise

Procuring an adapted version of Japan's *Sōryū*-class diesel-electric submarine provides an attractive alternative, particularly if combined with U.S. expertise in undersea sensors. The *Sōryū*-class is a large diesel-electric submarine, giving it the expanded payload and longer endurance necessary for Australia's large area of operations. In addition, the *Sōryū*-class's reliability, quietness, and advanced sensor systems provide a substantial capability increase over the *Collins*-class. As an alternative, Australia might seek to procure only the *Sōryū*-class's drive system, which proved to be the most difficult engineering challenge on the *Collins*-class.

Australian leaders have publicly discussed the potential to cooperate with Japan on the Future Submarine. These negotiations are Japan's first major foray into export of defense goods, and they would provide a model for how Japan might work with regional partners to build a more federated model of defense. U.S. cooperation will also be vital, particularly because the Future Submarine may employ a weapons suite of U.S. origin. The U.S. Navy can also play a major role in facilitating cooperation given its extensive experience in joint development programs. Although this issue is politically sensitive, such cooperation could allow Australia to procure a more capable submarine at a lower cost, improving not only its capabilities but those of the U.S.-Australia alliance.

# Missile Defense

As cruise and ballistic missiles proliferate, Asian states face a growing need for air and missile defenses. Most notably, China and North Korea are developing ballistic missiles, including nuclear-armed systems with the range to strike the cities and military assets of states throughout the region. China's missile forces present a particularly difficult challenge to regional militaries and U.S. forward-deployed forces. In addition, many regional players are acquiring long-range cruise missiles in greater numbers, including China, India, Russia, South Korea, and Taiwan. To respond to these threats, many regional states are seeking advanced missile defenses.

The United States already has a number of air and missile defense systems used by many in Asia. The Theater High Altitude Area Defense (THAAD) system provides capability against long-range ballistic missiles and has been deployed to Guam and may be deployed to South Korea. The Patriot surface-to-air missile system (which is coproduced by Japan) is used to protect many regional militaries against air and missile threats, including Taiwan and South Korea. In addition, several regional navies operate the Aegis Combat System, including Japan, South Korea, and Australia. Designed to be launched from Aegis-capable ships, the Standard Missile-3 (SM-3) Block IIA has been codeveloped by the United States and Japan. This is one of the most notable multinational efforts to develop advanced missile defenses. Aegis-ashore may also bolster regional missile defense capabilities, particularly if sensors can be integrated among states such as Taiwan, Japan, and South Korea. India has also tested its own ballistic missile intercept system with exoatmospheric and endoatmospheric intercept capability. In addition to these U.S.-fielded missile defense systems, some countries are looking into the possibility of acquiring foreign radars and interceptors, such as the codeveloped U.S.-Israel Arrow missile defense system.

Efforts to develop kinetic interceptors are vital to bolstering U.S., ally, and partner missile defense capabilities. Existing air and missile defenses may be able to intercept many missiles headed for critical assets and infrastructure; the PAC-3 Missile Segment Enhancement and extending the range of THAAD will further increase intercept capabilities. In the long term, however, these systems have some serious drawbacks. First, kinetic interceptors require advanced sensors and propulsion systems, often making the interceptors more costly than the missiles they are designed to strike. Second, because kinetic interceptors are expensive, they can usually be overwhelmed if the attacker sends a large salvo of incoming missiles. As a result, most kinetic intercept systems make defenders accept disadvantageous cost-exchange ratios in which they must spend more money to defend themselves than the attacker must spend to threaten them. Furthermore, such highly advanced missile defense capabilities tend to be available only to the most well-funded and technically capable regional militaries. Most states, particularly those in Southeast Asia, must focus instead on more limited air defense systems. Even advanced militaries face the challenge of developing cost-effective interceptors.

For these reasons, innovative approaches, such as railguns and directed energy defenses, will be critical in the long term. The United States, China, and other advanced militaries are likely to be the first movers on these technologies. Development of these defensive systems is likely to be costly for the United States and its allies, driving cooperation to lower costs while maximizing capability.

**Illustrative Example: Cooperation on Railgun and Directed Energy Defenses**

Most existing missile defense systems put the defender at a cost disadvantage. The cost of kinetic interceptors is prohibitive for many smaller states and presents a cost-exchange asymmetry for all states. These are serious challenges, but there are emerging technologies that could help the United States and its allies and partners to defend themselves. Foremost among these technologies are railgun and directed energy systems.

Railguns, currently being tested for both deployment at sea and on land, are potentially attractive because they use stored electricity to propel metal slugs fast enough to kinetically kill aircraft and missiles without requiring advanced warheads. Currently metal slugs are expensive, but costs should decrease as the volume of slugs produced increases. It is even possible that the United States could defray some of these costs by working with Asian partners to more cheaply produce railgun slugs.

Directed energy systems are in development by the United States and other regional militaries, such as China, Japan, and India. These systems store electricity, but rather than using it to propel metallic slugs, they focus beams of energy on their targets. These systems are potentially advantageous, particularly for large surface ships (such as nuclear-powered aircraft carriers) and ground-based locations that have the capacity to generate and store large amounts of electrical energy. If energy storage systems and beam-focusing technologies can be mastered, directed energy holds the possibility of nearly unlimited magazine depth, providing a cost-effective means of air and missile defense.

Given the critical importance of air and missile defenses for the viability of U.S. forward-deployed forces and the rapid advance of China's own anti-access and area denial capabilities, the United States is likely to continue to devote substantial funding to these advanced technologies. The United States should work with regional allies and partners to share some of the development and deployment costs of these air and missile defenses, to potentially include joint production of energy storage systems and railgun ammunition. Given partner expertise in battery technology and the growing threat from ballistic and cruise missiles, such cooperation seems both natural and urgent.

## Cybersecurity

Although cyber threats have multiplied in recent years, many Asian states continue to have limited capability for operations in cyberspace. Driven by the growth of cyber capabilities in many regional countries, particularly China and North Korea, there is a natural desire for cooperative efforts to defend both government and commercial networks. Although most countries in Asia have increased spending on cyber capabilities, many national cyber-defense efforts are understaffed, to say nothing of the need to develop common rules and norms for regulating cyber capabilities. Here again a more federated approach could increase capability while controlling costs.

In 2011, the United States issued an International Strategy for Cyberspace that supported principles such as upholding fundamental freedom, respect for property, valuing privacy, protection from crime, and the right of self-defense.[3] In addition to these principles, the administration encouraged norms of global interoperability, network stability, reliable access, multistakeholder governance, and cybersecurity due diligence. Operationalizing this strategy requires a concerted effort to engage regional states, particularly the quickly growing and deeply interconnected network of Asian economies.

Many cyber capabilities remain sensitive, limiting the information that is publicly available about ongoing efforts. In thinking about these operations, however, former U.S. deputy secretary of defense Ash Carter has noted that the United States has identified three separate but related missions in cyberspace.[4] The first mission is to defend U.S. national networks against strategic cyber attacks. The second mission is to protect Department of Defense information networks against cyber attacks. The third mission is to support combatant commands in carrying out combat operations. These three missions are typically termed *the national mission*, *cyber protection mission*, and *combat mission*, respectively. These three missions apply not only to the United States but to Asian allies and partners as well. Multinational efforts in each area are needed to address emerging challenges and opportunities.

National mission forces, those that protect national networks against attack, are necessary to protect against intrusions from both state and non-state actors. National infrastructures are under constant threat from cyber attacks targeted against everything from major financial institutions to power and energy infrastructure. Cyber espionage, to which all states are vulnerable, also poses a threat to corporations. These dangers require investments by both the public and private sectors, but the strategic nature of the threat against national infrastructure is convincing most states that a government-funded capability is needed to help protect national networks against attack. This is a shared threat, which is prime for cooperative efforts.

3. "International Strategy for Cyberspace," White House, 2011, http://www.whitehouse.gov/sites/default/files/rss_viewer/international_strategy_for_cyberspace.pdf.

4. "Remarks by Deputy Secretary of Defense Carter at the Aspen Security Forum at Aspen, Colorado," U.S. Department of Defense, July 18, 2013, http://www.defense.gov/Transcripts/Transcript.aspx?TranscriptID=5277.

Cyber protection forces carry out a similar mission to national mission forces, except that instead of focusing on the nation as a whole, they are devoted only to the security and operation of critical government (typically military) systems. U.S. government networks are under constant threat, which means that U.S. cyber protection experts are deeply experienced in defending their networks. Many U.S. allies and partners, on the other hand, face fewer threats, particularly during peacetime. As a result, U.S. expertise in cyber protection is vital to many regional partners who are likely to face a severe cyber threat during wartime but enjoy a more pacific cyber environment in peacetime.

Finally, combat mission forces are those that support military operations, to include offensive cyber operations. Although there is great sensitivity surrounding offensive cyber operations, their proliferation makes coordination of such operations necessary and important. In particular, China's frequent and well-documented use of offensive cyber operations means that other regional states will have to develop their own countervailing capabilities or risk failure in deterrence and defense against cyber attacks. Although information and intelligence sharing surrounding offensive cyber capabilities is likely to be sensitive, here again the United States has critical capabilities and expertise that are likely to be sought throughout Asia.

**Illustrative Example: Combined Cyber Exercises**

Asian leaders acknowledge that improved cyber capabilities will be critical to the region's security and prosperity. Few efforts exist, however, to develop multilateral training and exercising for cyber contingencies. As part of its efforts to develop and exercise new capabilities, operational concepts, and contingency plans for assuring access in Asia, the United States should lead a series of combined exercises with Asian allies and partners that feature cybersecurity as a central element. Such efforts could both publicize the importance of cybersecurity and drive future investments in the capabilities required for effective operations.

Cyber operations should be seen as a central element of U.S. efforts to assure access throughout Asia. U.S.-led war games could help to integrate regional efforts, deepen understanding of U.S. warfighting capabilities, and act as a deterrent against states that might seek to deny Asian allies and partners the use of the cyber domain in a conflict. Most importantly, such exercises would help regional partners to identify weaknesses in their cyber capabilities and to work together to address these deficiencies. Exercises could begin with national mission forces, then expand to include cyber protection forces, and finally address cyber combat forces and the need to de-conflict offensive actions in cyberspace.

# 5 Conclusion

As Asian security challenges grow, the concept of federated defense is likely to become more central to regional security and prosperity. Rising threats will require that regional states work together to build capability and capacity across the spectrum of potential mission areas. If U.S., ally, and partner defense budgets remain under pressure, then maximizing the value of defense spending at home and abroad will increasingly attract the attention of both policymakers and the public.

The U.S. rebalance to Asia provides an opportunity to make federated defense a core component of U.S. regional strategy. Since the rebalance was announced in 2011, the U.S. military has enhanced its regional engagement and presence by signing new access agreements, realigning U.S. military posture, rotationally deploying U.S. units, forward deploying advanced systems, and bolstering training and exercising with regional allies and partners. Federated defense could link these initiatives together and provide a framework for deepening existing efforts, all while providing new opportunities for cooperation. As various countries expand and deepen capability and capacity in certain areas, regional states will need to reassess existing roles and missions. The proposals set forth here can form the basis for these efforts.

Implementing federated approaches will require close cooperation not only between the United States and its foreign allies and partners but also within various parts of their governments and defense industries. Executive leadership will be vital if federated approaches are to gain traction, particularly between the Office of the Secretary of Defense, the military services, the U.S. Pacific Command, the intelligence community, the Department of State, and the National Security Council. Congressional support and funding will be critical to enable and sustain these efforts. Defense industry's embrace of federated initiatives will be necessary to realize the desired technological innovations and cost savings. Moreover, outreach to the general public will be required to provide the necessary understanding of and support for such cooperative efforts.

The project codirectors envision this report as the beginning of an effort to study and apply federated initiatives to Asia. This report demonstrates the need for and potential of federated approaches in Asia. Although this study highlights initiatives in six specific areas—humanitarian assistance and disaster relief, information and intelligence sharing, maritime security, undersea warfare, missile defense, and cybersecurity—federated defense applies across the range of security missions. From combating terrorism in South and Southeast Asia to countering anti-access and area denial threats in Northeast Asia, the need for federated approaches is growing. This report provides a road map for how these types of initiatives might be approached, but additional follow-on efforts will focus on the regional and functional changes required to build a federated defense.

Despite significant challenges, the opportunities for federated defense cooperation are substantial and the need is great. Under pressure to "do more with less," federated defense will be central to regional efforts to grow regional capacity and capability in the years ahead. Yet federated defense cooperation is vital regardless of budget environments; the increasing complexity and connectedness of regional security challenges require coordinated efforts. Federated approaches are vital to developing and integrating the region's security capabilities, thereby reinforcing security and prosperity not only within Asia but beyond.

ASEAN foreign ministers and Secretary of State Hillary Clinton pose at the 18th ASEAN Regional Forum Retreat Session in Bali, Indonesia, on July 23, 2011. Participant countries include Mongolia, Myanmar, New Zealand, Pakistan, Papua New Guinea, Philippines, Russia, Republic of Korea, Singapore, Sri Lanka, Thailand, Timor-Leste, United States, Vietnam, DPRK, European Union, India, Japan, Laos, Malaysia, Indonesia, Cambodia, Australia, Bangladesh, Brunei, Canada, China, and ASEAN. Photo by Erik Kurniawan.

# About the Authors

**Michael J. Green** is senior vice president for Asia and Japan Chair at CSIS and an associate professor at the Edmund A. Walsh School of Foreign Service at Georgetown University. He served on the staff of the National Security Council (NSC) from 2001 through 2005, first as director for Asian affairs, with responsibility for Japan, Korea, Australia, and New Zealand, and then as special assistant to the president for national security affairs and senior director for Asia, with responsibility for East Asia and South Asia. Before joining the NSC staff, he was senior fellow for East Asian security at the Council on Foreign Relations, director of the Edwin O. Reischauer Center and the Foreign Policy Institute, and an assistant professor at the Paul H. Nitze School of Advanced International Studies (SAIS) at Johns Hopkins University, research staff member at the Institute for Defense Analyses, and senior adviser on Asia in the Office of the Secretary of Defense. He also worked in Japan on the staff of a member of the National Diet.

Dr. Green is also currently a nonresident fellow at the Lowy Institute in Sydney, Australia, and a distinguished scholar at the Rebuild Japan Initiative Foundation in Tokyo. He is a member of the Council on Foreign Relations, the Aspen Strategy Group, the America Australia Leadership Dialogue, the advisory board of the Center for a New American Security, and the editorial boards of the *Washington Quarterly* and the *Journal of Unification Studies* in Korea. He is also an associate of the U.S. Intelligence Community. Dr. Green has authored numerous books and articles on East Asian security. His current research includes a book project on the history of U.S. strategy in Asia; a survey of elite views of norms, power, and regional institutions in Asia; and a monograph on Japanese strategic culture. He received his master's and doctoral degrees from SAIS and did additional graduate and postgraduate research at Tokyo University and the Massachusetts Institute of Technology. He received his bachelor's degree in history from Kenyon College with highest honors.

**Kathleen H. Hicks** is senior vice president, Henry A. Kissinger Chair, and director of the International Security Program at CSIS. She is a frequent writer and lecturer on U.S. foreign and security policy; defense strategy, forces, and budget; and strategic futures. From 2009 to July 2013, Dr. Hicks served as a senior official in the Department of Defense. Confirmed in May 2012 as principal deputy undersecretary of defense for policy, she was responsible for advising the secretary of defense on global and regional defense policy and strategy pertaining to such areas as the Asia-Pacific and Persian Gulf regions, Syria, and Europe. She also served as deputy under secretary of defense for strategy, plans, and forces, leading the development of the 2012 Defense Strategic Guidance and the 2010

Quadrennial Defense Review and crafting guidance for future force capabilities, overseas military posture, and contingency and theater campaign plans.

From 2006 to early 2009, Dr. Hicks served as a senior fellow at CSIS, leading a variety of research projects in the national security field, including as part of the Center's Beyond Goldwater-Nichols series. She was a career civil servant in the Office of the Secretary of Defense from 1993 to 2006, serving in a variety of capacities and rising from Presidential Management Intern to the Senior Executive Service. Dr. Hicks has received numerous recognitions for her service in the Department of Defense (DOD), including distinguished awards from Secretaries of Defense Chuck Hagel, Leon Panetta, and Robert Gates and from chairman of the Joint Chiefs of Staff General Martin Dempsey and the 2011 DOD Senior Professional Women's Association Excellence in Leadership Award. She holds a Ph.D. in political science from the Massachusetts Institute of Technology, an M.A. from the University of Maryland's School of Public Affairs, and an A.B. magna cum laude and Phi Beta Kappa from Mount Holyoke College. She previously served as an ex officio member of the Board of Directors for the U.S. Institute of Peace and currently serves on the Board of Advisers for the Truman National Security Project. Dr. Hicks is also a member of the Council on Foreign Relations and an adjunct with the RAND Corporation.

**Zack Cooper** is a fellow with the Japan Chair at the Center for Strategic and International Studies, where he focuses on Asian security issues. Mr. Cooper is also a doctoral candidate in security studies at Princeton University's Woodrow Wilson School. Prior to joining CSIS, Mr. Cooper worked as a research fellow at the Center for Strategic and Budgetary Assessments. He previously served on the White House staff as assistant to the deputy national security adviser for combating terrorism. He also worked as a civil servant in the Pentagon, first as a foreign affairs specialist and then as a special assistant to the principal deputy undersecretary of defense for policy. Mr. Cooper received a BA from Stanford University and an MPA from Princeton University, where he served as the director of strategic education for the school's Center for International Security Studies.

www.ingramcontent.com/pod-product-compliance
Ingram Content Group UK Ltd.
Pitfield, Milton Keynes, MK11 3LW, UK
UKHW051418150926
475561UK00005B/234

9 781442 240452